Siddhesh Pandurang Narvekar
Prathamesh Shivaji Dhumal
Shankar Nitin Ganachari

ECO-FRIENDLY CONCRETE USING BACTERIA COATED PLASTIC AGGREGATES

Siddhesh Pandurang Narvekar
Prathamesh Shivaji Dhumal
Shankar Nitin Ganachari

ECO-FRIENDLY CONCRETE USING BACTERIA COATED PLASTIC AGGREGATES

Sustainable Construction Material

LAP LAMBERT Academic Publishing

Imprint

Cover image: www.ingimage.com

Publisher:
LAP LAMBERT Academic Publishing
is a trademark of
Dodo Books Indian Ocean Ltd. and OmniScriptum S.R.L publishing group

120 High Road, East Finchley, London, N2 9ED, United Kingdom
Str. Armeneasca 28/1, office 1, Chisinau MD-2012, Republic of Moldova, Europe
Managing Directors: Ieva Konstantinova, Victoria Ursu
info@omniscriptum.com

Printed at: see last page
ISBN: 978-620-9-10524-1

ECO-FRIENDLY CONCRETE USING BACTERIA COATED PLASTIC AGGREGATES

Siddhesh P. Narvekar[1]

1 Assistant Professor, Department Of Civil Engineering,
Sant Gajanan Maharaj College of Engineering, Mahagaon, Kolhapur, MS, India

narvekar.siddhesh@gmail.com

09011323796

Prathamesh Shivaji Dhumal [2], Shankar Nitin Ganachari [3], Kiran Rajendra Kori [4]

2,3,4 Research Students, Department Of Civil Engineering,
Sant Gajanan Maharaj College of Engineering, Mahagaon, Kolhapur, MS, India

prathameshdhumal31@gmail.com; ganacharishankar30@gmail.com;

kirankori18@gmail.com

08888927982,07057206902,09370421265

ABSTRACT

The rising consumption of plastics has resulted in large quantities of non-biodegradable waste, creating severe environmental challenges. Recycling waste plastics into concrete as coarse aggregates provides a sustainable alternative, but it often leads to reduced strength due to poor bonding between plastic and cement paste. This study investigates the use of *Bacillus subtilis* bacteria coating on waste plastic aggregates to improve concrete performance. *Bacillus subtilis* spores, known for their ability to precipitate calcium carbonate ($CaCO_3$) under favorable conditions, were used to bio-coat shredded plastic aggregates. Concrete specimens were prepared with varying percentages of waste plastic aggregates (5%, 10%, 15%,20%) with and without bacterial coating. Tests were conducted on compressive strength, and crack-healing efficiency. Preliminary results indicate that bacterial coating enhances bonding, improves compressive strength, reduces permeability, and provides self-healing capabilities. This novel approach integrates plastic recycling and bio-concrete technology, offering a pathway toward sustainable construction

***Keywords*- Compressive strength of concrete, Waste plastic, Bacillus subtilis bacteria**

INDEX

INTRODUCTION

Concrete is the backbone of modern infrastructure, but its durability is challenged by cracking and permeability issues. Simultaneously, plastic waste is increasing globally, leading to serious environmental hazards. Incorporating shredded plastic into concrete as aggregate reduces the burden on landfills and conserves natural resources. However, plastics have low surface roughness and hydrophobicity, leading to weak interfacial bonding and reduced compressive strength.

To overcome this limitation, we have tried biological approaches such as *Bacillus subtilis*–based bio-concrete. *Bacillus subtilis* is a spore-forming bacterium that survives in the alkaline concrete environment and precipitates calcium carbonate, which can fill pores and cracks. Applying this bacterial coating to waste plastic aggregates may improve adhesion, mitigate strength loss, and provide self-healing capacity.

This study explores the feasibility of using bacterial-coated waste plastic aggregates in concrete and evaluates their effect on mechanical and durability properties. We perform various tests in laboratory at Sant Gajanan Maharaj College of Engineering, Mahagaon, Gadhinglaj for demonstrating the accuracy and applicability of the proposed methodology.

METHODOLOGY

Experimental Program:

1. **Plastic Aggregate Preparation:** Collection of waste plastic, cleaning, shredding, and sieving to coarse aggregate size.
2. **Cultivation of Bacteria:** Cultivation of bacteria's in culture medium
3. **Bacterial Coating:** Immersing plastic aggregates in bacterial suspension with carrier medium, then drying and curing.
4. **Mix Proportioning:**
    - Control Mix (0% plastic).
    - 5%, 10%, 15% plastic aggregate replacement (with and without bacterial coating).
5. **Casting & Curing:** Casting of concrete cubes (150×150×150 mm) and curing for 7, 14, 28 days.
6. **Testing:**
    - Workability (slump test).
    - Compressive strength (7, 14, 28 days).
    - Crack healing observation under moist curing (microscopy)

Materials:

- Cement: OPC 43/53 grade.
- Fine aggregate: River sand.
- Coarse aggregate: Natural aggregates + shredded waste plastic aggregates.
- Bacteria: *Bacillus subtilis* cultured in nutrient broth.
- Nutrient carrier: Calcium lactate / sodium alginate.

LITERATURE REVIEW

Gehad A. M. Metwally, Mohamed Mahdy, Ahmed H. A. El-Raheen,
'Performance of Bio Concrete by using Bacillus Pasteurli Bacteria'
Civil Engineering Journal, Vol. 6, No. 8, August 2020.

INTRODUCTION

The construction industry in the world has recently testified an enormous development. Concrete as a construction material received widespread use throughout the world during the 20th century. The steady production of Pre-mixed concrete is not only one of those significant indicators of concrete technical success but also totally increases in performance, but we face several novel challenges. The major prevalent issue that there is an increased hazard of rifting owing to Offloading circumstances back to poor Concrete tensile capacity, like cracking, calorific rifting also chemical Responses rifting Because of steel abrasion, Response of alkali-silica, sulfate raid. Cracking increases the likelihood of the hazardous object penetrating the concrete, which negatively affects the integrity of the component since they provide an easy path for the transport of liquids and gases that potentially contain harmful substances. When micro-cracks germinate and strike the reinforcement, not just the concrete itself can be threatened, but also the reinforcement will be rusted when exposed to water, oxygen, and possibly carbon dioxide and chlorides. Micro-cracks are, therefore, precursors to structural failure. Commonly cracks are remedied manually, which is unconvincing since cracks are often difficult to identify, and repair and maintenance costs are high. Self-healing of cracked concrete would, therefore, be extraordinarily beneficial, and research into self-healing concrete has been broadly performed out

While in this study the optimal blending content proportion was identified (10SF20BC), which showed an increment in compressive strength and flexural strength compared to the control mixture to reach 79.16%, 50% respectively, also enhanced the microstructures of concrete with the concentration of 5×109 cells/ml and OD**600**. From another point of view, interesting changes in strength were recorded at an early age. Such finding might be due to the fact that urea, calcium and nutrient were able to find access in the porous concrete to reach the bacteria in order to implement the target biochemical reaction at an early age.

CONCLUSION

Bacteria considered being alkali immune, which grows in inherent surroundings distinguished by comparatively elevated pH, varies from 10 to 11. Furthermore, this mentioned progeny may generate germs that restful the cells with robust cell walls that shield cells from severe surrounding mechanical and chemical Fatigue. As a consequence, bacillus pasteurii can obtain the capacity to resist elevated inner concrete pH rates (12 to 13 for conventional concrete) and continue to survive for a longer period. This is assumed to know that the bacteria immobilized germs can able to fill rifts through Biological-mineral development following they have been resurrected by water and Growth of nutrients reaching newly formed rifts. Compressive strength test results observed improvement by utilizing the mixture of SCMs with a self-healing agent (Bacillus pasteurii). The compressive strength development relied on a type of utilized materials, utilized materials mixing ratio and water to binder ratio, MIV results showed an increment in compressive strength compared to the control mixture to reach 79.16% (w/c= 0.40, Fcu=250 kg/cm2), where MV achieved enhancement to 11.11% related to its control blend (w/c=0.35, Fcu=400 kg/cm2), while the high strength concrete obtained raising to 7.14% related to control mixture(w/c=0.302, Fcu=800 kg/cm2). Flexural strength was also improved within the range of (15 to 20) % replacement of blended water . MV results showed an increment compared to the control mixture to reach 400.32% (w/c=0.40, Fcu =250 kg/cm2), where MIV achieved enhancement to 142.82% related to its control blend (w/c=0.40, Fcu=400 kg/cm2), while the high strength concrete mixture obtained rising in flexural strength to 53.14% related to control mixture (w/c=0.302, Fcu=800 kg/cm2). The optimum mixing material ratio was 10SF20BC, which displayed the highest percentage of calcium carbonate precipitations 9.49% of hweight of mixtures element in the range of 5×109 cells/ml to the mixing water. The best crack width could be healed was varied between 0.1-0.60 mm, and after that, the concentration of cells in bacteria should be increased as shown in MIV.

Azad Khajuria and Puneet Sharma

'Use of Plastic Aggregates in Concrete'

International Journal of Innovative Technology and Exploring Engineering (IJITEE) ISSN: 2278-3075 Volume -9 issue -1, November 2019

INTRODUCTION

Concrete is the mostly used man made material used in construction industry and is the second after water as the most utilized thing on the Earth. In simple words it is defined as a mixture of four ingredients as coarse aggregates that form the largest proportion of the mix, fine aggregates such as sand that act as filler material in the voids, binding material such as lime or Portland cement that binds these material together and water that reacts with binding material. The mixing of these four materials gives us a paste that is called as matrix. At this stage it is called as fresh concrete or green concrete and get hardened like a stone, as the water reacts with binding material. This reaction is called as hydration of concrete. In fresh state concrete can be casted into any desired shape by placing it in forms. This property of concrete help in using the concrete in most efficient manner. Plastic needs no introduction as it is the widely used material now a days on our Earth. Due to its properties like strength, durability and easy processing it can be used for many purposes. Studies shows that plastic is nearly inert that is it get very less affected by the chemicals and have higher durability. Disposal of plastic waste is a huge problem as due to absence of organic compounds, it is non decomposable material and proves to be a threat to our environment as it has many health hazards. As decomposition of plastic is a serious problem as it takes very long time and adversely affection the environment in many ways. So we can use it in construction, where we need life of structure to be improved and use of waste plastic after small processing can help us to reduce the waste in the environment which is new motto of civil engineering.

CONCLUSIONS

Plastic was added to concrete in replacement of coarse aggregates by proportion of 0, 2.5, 5, 7.5 and 10%. On the basis of the results from the present study, following conclusions were drawn i. The material used in the experiments is good and workable. ii. The admixture used in the experiments gave the great impact on the strength of concrete. iii. The specific gravity of plastic was lesser than that of aggregates. iv. While testing the flexural strength of the beam, it is seen that beam failed in between the loading span between its two supports and hence formula that we used is 3PL/4bd2. v. It was observed while experiment that the compressive strength of concrete initially increases at 2.5% PCA but further addition of PCA shows reduction in strength. vi. The optimum compressive strength is obtained at 2.5% PCA. vii. The tensile strength of cylinder shows better result as compared to other strengths. viii. Flexural Strength shows similar result to that of compressive strength

M Mahesh, B Venkat Narsimha Rao, CH. Satya Sri

'Re-use of Polyethylene Plastic Waste in Concrete'

International Journal of Engineering Development and Research; Volume 4, Issue 4 ISSN:2321-9939.

INTRODUCTION

Due to rapid industrialization and urbanization in the country lots of infrastructure developments are taking place. This rapid development led to the acute shortage of construction materials, increased dumping of waste materials. Hence to overcome the above said problems waste products should be employed as a construction material. Fine aggregate used in concrete is replaced partially by pulverized PET bottles in known percentages and properties are tested, the optimum percentage at which higher strength is obtained is calculated. Considerable researches were carried out in some countries like USA, UK on this topic however there have been very limited studies on plastics in India

CONCLUSION

Waste plastic can be effectively re-used without affecting the mechanical properties considerably (5-10%).

With increase in the percentage of plastic there has been a sudden decrease in early strength but the strength developed to the value as that of the conventional M25 concrete when 28 day tests were performed.

It is observed that for more percentage addition of plastics i..e 6% in the present case, the 7 day strength has been decreased by --- when compared with conventional concrete. For less percentage addition (2-4%) of plastic, there is no considerable variation in 7 day, 14 day, and 28 day compression strength and split tensile strength.

Since Specific gravity of waste plastic is less than that of fine aggregate, thus self weight of concrete reduces, thus it reduces the weight of the structure/structural component as a whole.

Concrete with plastic waste can be used for less important works where concrete is not going to bear more loads.

Nain, Nidhi, R. Surabhi, N.V. Yathish, V. Krishnamurthy, T. Deepa, and Seema Tharannum.

'Enhancement in Strength Parameters of Concrete by Application of Bacillus Bacteria'

Construction and Building Materials 202 (March 2019

INTRODUCTION

Concrete has proven to be useful since 6500 BCE and has now become the second most widely used element after water On the contrary, concrete possesses lower tensile strength and less ductility due to which it corroborates vulnerability towards the generation of cracks. The production of cracks not only lowers the strength of concrete but also gives access to deleterious substances. Besides this failure of concrete, micro-cracks are inherently present in concrete. Both the prior is well-known to result in debilitation of structures.

Referring to a series of research carried out around the world, various methodologies have been carried out sporadically with the main motive to overcome deficiencies of cement concrete. The most common surface treatments use organic polymers (epoxy, siloxane, acrylics and polyurethanes). These methods used for managing micro-cracks, have a certain degree of toxicity and also possesses certain drawbacks that make them not completely efficient. A recent series of research shows a distinguish use of microbial mineral precipitation resulting from metabolic activities of favorable microorganisms in improving the overall behavior of concrete. The process takes place near to or at any distance away within the concrete. This process of bacterial application to enhance concrete property is also termed as Bio mineralization. In the process, urease enzyme catalyzes the hydrolysis of urea into ammonia and carbonate, this carbonate reacts with calcium and thereby forming 'calcium carbonate precipitate'

The past experimental journal shows the increase in the concrete property by use of various bacteria of genus Bacillus, genus Sporosarcina, and E. coli. In coordination with the prior researches, this journal proves experimentally the use of bacteria of genus Bacillus (Bacillus subtilis & Bacillus megaterium) in enhancing concrete properties and also compares the suitability of consortia of both the Bacterium for the same

CONCLUSION

The following conclusions can be drawn based on the present study-

The plate count test after curing proves the viability of bacteria in concrete and its ability to withstand the harsh environment.

Subtilis shows a percentage increase of 14.3% in compressive strength and 25.3% increase in tensile strength compared to the conventional specimen.

Megaterium shows a percentage increase of 22.5% in compressive strength and 18.49% increase in tensile strength compared to the conventional specimen.

Consortia of bacteria show a percentage increase of 15.8% in compressive strength and 19.58% increase in tensile strength compared to the conventional specimen.

The slower strength gains of megaterium depicts that megaterium has a slower ability to obtain the precipitation in comparison with B. subtilis even after it's proved viability after 7 days.

Consortia show a percentage increase of 1.3% in compressive strength compared to B. subtilis and a percentage decrease of 5.8% in same when compared to B. megaterium.

Consortia show a percentage decrease of 4.8% in tensile strength compared to B. subtilis and a percentage increase of 1.03% in same when compared to B. megaterium.

SEM images and EDAX result analysis gives a strong proof of the presence of calcite precipitation.

The study conclusively indicated that the microorganism demonstrated a positive role in enhancing the strength of concrete which is due to filling up of voids by calcite precipitation in the concrete specimen. Hence, there is a possibility of self-healing of microcracks for which further studies are required.

Text Book of Concrete Technology, Theory and practice by M.S. Shetty.

Cement concrete is one of the seemingly simple but actually complex materials. Many of its complex behaviors are yet to be identified to employ this material advantageously and economically. The behavior of concrete with respect to long-term drying shrinkage, creep, fatigue, morphology of gel structure. Bond, fracture mechanism and polymer modified concrete, fibrous concrete are some of the areas of active research in order to have a deeper understanding of the complex behavior of these materials.

The book deals with several aspects of concrete technology and also covers the latest developments taken place in India and abroad. The coverage is comprehensive and complete. The properties of constituent's materials of concrete have been explained very lucidly in the text.

INTRODUCTION TO CEMENT

The term cement is commonly used to refer to powdered materials which develop strong adhesive qualities when combined with water. Cement is a binding material used in construction to bond materials like sand and aggregates. It is primarily made of limestone, clay, and other additives, which undergo a heating process to form a fine powder.

Cement was first invented by the Egyptians. Cement was later reinvented by the Greeks and the Babylonians who made their mortar out of lime. Later, the Romans produced cement from pozzolana, an ash found in all of the volcanic areas of Italy, by mixing the ash with lime.

Cement is a fine grayish powder which, when mixed with water, forms a thick paste. When this paste is mixed with sand and gravel and allowed to dry it is called concrete.

About ninety-nine percent of all cement used today is Portland cement. The name Portland cement is not a brand name. This name was given to the cement by Joseph Aspdin of Leeds, England who obtained a patent for his product in 1824. The concrete made from the cement resembled the color of the natural limestone quarried on the Isle of Portland in the English Channel. The balance of cement used today consists of masonry cement, which is fifty percent Portland cement and fifty percent ground lime rock.

The first cement manufactured in the United States was produced in 1871 by David Saylor of Coplay, Pennsylvania.

There are two types of raw materials which are combined to make cement:

- Lime-containing materials, such as limestone, marble, oyster shells, marl, chalk, etc.
- Clay and clay-like materials, such as shale, slag from blast furnaces, bauxite, iron ore, silica, sand, etc

It takes approximately 3,400 lbs. of raw materials to make one ton (2,000 lbs.) of Portland cement. The mixture of materials is finely ground in a raw mill. The resultant raw mix is burned in a rotary kiln at temperatures around 4482 degrees Celsius to form clinker. The clinker nodules are then ground with about 3 % gypsum to produce cement with a fineness typically of less than 90 micrometers.

Types of Cement

1) Ordinary Portland Cement (OPC)
   - Most commonly used cement
   - Available in Grades: 33, 43, 53
   - Used in general construction, buildings, bridges
2) Portland Pozzolana Cement (PPC)
   - Contains fly ash for better durability
   - Used in marine structures, dams, and bridges
3) Rapid Hardening Cement
   - Gains strength quickly
   - Used for fast-track projects like roads

4) Sulfate-Resistant Cement
   - Prevents damage from sulfate-rich environments
   - Used in sewage treatment plants
5) Low Heat Cement
   - Generates less heat during hydration
   - Suitable for massive concrete structures like dams
6) White Cement
   - Made from pure limestone and clay
   - Used for decorative purposes and tile fixing
7) Quick Setting Cement
   - Sets faster than OPC
   - Used in underwater construction
8) High Alumina Cement
   - High heat resistance
   - Used in refractory applications
9) Green Cement
   - Eco-friendly with reduced carbon emissions
   - Made using industrial waste like slag and fly ash

Cement Testing in Laboratory

1. Fineness Test – Determines the particle size
2. Soundness Test – Checks volume stability
3. Consistency Test – Measures water content
4. Setting Time Test – Determines initial and final setting time
5. Compressive Strength Test – Measures strength of cement

PROPERTIES OF CEMENT

1. Fineness
2. Soundness
3. Specific Gravity
4. Heat of hydration
5. Bulk density
6. Consistency
7. Setting time
8. Compressive strength

CONCRETE TESTING IN LABORATORY

1. Workability (slump test).
2. Compressive strength (7, 14, 28 days).
3. Water absorption.
4. Rapid chloride permeability test.
5. Crack healing observation under moist curing (microscopy)

1. WORKABILITY TEST

A. Slump Test

Suitable for medium to high workability concrete (not very stiff or very fluid).

Apparatus:

- Slump cone (300 mm high, 200 mm bottom dia, 100 mm top dia)
- Tamping rod (16 mm dia, 600 mm long, rounded end)
- Non-absorbent base plate

Procedure:

1. Clean and place the slump cone on a leveled, non-absorbent surface.

2. Fill the cone in 3 equal layers of fresh concrete.
3. Compact each layer with 25 strokes of the tamping rod.
4. Strike off excess concrete and lift the cone vertically upward in 5–10 seconds.
5. Measure the subsidence (slump) of concrete from cone height.

Types of Slump:

- True slump → Uniform settlement (ideal)
- Shear slump → One side drops (not desirable)
- Collapse slump → Concrete too wet (high workability)

Result Expression:

Slump (mm) = Height of cone – Height of slumped concrete

Typical Values: 25–125 mm depending on application.

B. Compaction Factor Test

Suitable for stiff concrete (low workability).

Apparatus:

- Compaction factor apparatus (two conical hoppers and one cylindrical container)
- Tamping rod
- Balance (for weighing concrete)

Procedure:

1. Fill the upper hopper with fresh concrete (without compaction).
2. Open trap door → concrete falls into second hopper.
3. Open second trap door → concrete falls into the cylindrical container.
4. Strike off excess and weigh the concrete in cylinder (partially compacted weight).

5. Now fill cylinder with same concrete in layers, compacting each fully. Weigh again (fully compacted weight).

Compaction Factor Formula:

CF = Partially compacted weight / Fully compacted weight CF

Typical Range: 0.7 (very low workability) to 0.95 (high workability).

C. Vee-Bee Consistometer Test

Suitable for very stiff concrete (workability too low for slump test).

Apparatus:

- Vee-Bee consistometer (cylindrical container, slump cone, transparent disc, vibrator table).

Procedure:

1. Place slump cone inside cylindrical container, fill and compact concrete in cone (as in slump test).
2. Remove cone → concrete remains in slump shape inside container.
3. Place transparent disc on top and switch on vibrator.
4. Note the time taken (in seconds) for concrete to remold fully into cylindrical shape.

Result Expression:

Vee-Bee time (s) → Lower time = higher workability

Typical Range: 5–30 seconds.

D. Flow Table Test

Suitable for very high workability concrete (flowing concrete, SCC).

Apparatus:

- Flow table (700 mm square, 76 mm thick, polished surface)
- Slump cone

Procedure:

1. Place slump cone at center of flow table and fill with concrete in layers (compact).
2. Lift cone vertically.
3. Raise table and drop it 15 times in 15 seconds.
4. Measure diameter of spread concrete in two directions at right angles.

Result Expression:

Flow (%) = (Spread diameter – Base diameter) / Base diameter

Typical Values: 50–150% (depending on SCC or high-workability needs).

2. COMPRESSIVE STRENGTH OF CONCRETE

It gives the characteristic strength of concrete, which is the basis for mix design and quality control.

Objective

To determine the load-carrying capacity of hardened concrete under axial compression.

Apparatus Required:

- Compression Testing Machine (CTM) (capacity ≥ 2000 kN, rate of loading ~140 kg/cm²/min).
- Cube moulds: 150 mm × 150 mm × 150 mm (standard size, IS code).
- Tamping rod (16 mm dia, 600 mm long).
- Vibrating table (optional, for compaction).
- Curing tank with clean water.

Procedure:

1. Preparation of Test Specimens

1. Clean and oil the cube moulds.

2. Fill with fresh concrete in 3 layers, each compacted by 25 strokes of tamping rod (or vibration).
3. Strike off excess concrete to get smooth top surface.
4. Store moulds at room temperature (20–27 °C) for 24 ± ½ hours.

2. Curing of Specimens

1. After 24 hours, remove the cubes from moulds.
2. Place them in a curing tank filled with clean water.
3. Maintain water temperature at 27 ± 2 °C.
4. Cure for required test ages (commonly 7 and 28 days).

3. Testing of Cubes

1. Remove cube from water and wipe off surface moisture.
2. Measure cube dimensions (for area calculation).
3. Place cube centrally in the CTM on its smooth faces.
4. Apply load uniformly at 140 $kg/cm^2/min$ until failure.
5. Note the maximum load at failure (P).

Calculations:

Compressive Strength (fc)} = {P} /{A}

Where:

- P = Maximum load applied at failure (N)
- A = Loaded area of cube (mm^2)

For a 150 mm cube:

A = 150 × 150 = 22500 mm2

Standard Values (as per IS 456:2000):

7-daystrength ≈ 65–70% of 28-day strength

14-day strength ≈ 85–90% of 28-day strength

Modes of Failure:

- Brittle failure with sudden cracking (normal concrete).

- Cone-shaped failure at top or bottom.
- Crushing failure along edges.

3. WATER ABSORPTION TEST

Objective:

- To determine the amount of water absorbed by concrete.
- It indicates porosity, durability, and quality of the material.

Apparatus:

1. Weighing balance (accuracy ±0.1 g).
2. Oven (capable of 105–110 °C).
3. Water tank or container for immersion.
4. Dry cloth for surface wiping.

Procedure:

1. Drying of Sample

1. Take the concrete cube/brick/aggregate sample.
2. Place it in a ventilated oven at 105–110 °C.
3. Dry until it achieves constant weight (two consecutive weights at 2-hour intervals differ by ≤0.2%).
4. Record this Oven-dry Weight (W_1).

2. Immersion in Water

1. Cool the dry specimen to room temperature.
2. Immerse completely in clean water at 27 ± 2 °C **for** 24 hours.

3. Surface Drying

1. Remove specimen from water after 24 hrs.
2. Wipe off surface water with a damp cloth (do not heat or sun-dry).

4. Weighing after Immersion

1. Immediately weigh the specimen.

2. Record this as Water-saturated Weight (W_2).

Calculation:

Water Absorption (%) = (W2−W1) / W1×100

Where:

- W1 = Oven-dry weight (gm)
- W2 = Water-saturated weight (gm)

Standard Values:

- Concrete cubes → ≤ 5–6% (as per IS 2185).

Importance:

- Lower water absorption = higher density, durability, and resistance to frost/chemical attack.
- Higher water absorption = porous, weaker, and less durable concrete/brick.

4. **RAPID CHLORIDE PERMEABILITY TEST (RCPT)**

Objective:

- To measure the resistance of concrete against chloride ion penetration.
- Chloride ions cause corrosion of reinforcement → reduction of service life.
- RCPT gives a quick indication of concrete durability in chloride-laden environments (marine, coastal, de-icing salts).

Apparatus Required:

- RCPT test cell (two-part cell with seals).
- DC power supply (capable of applying 60 V).
- Ammeter (0–500 mA range).
- Data logger (optional).
- Vacuum desiccator + vacuum pump.
- Oven (for drying).

- Measuring cylinder.
- NaCl solution (3% sodium chloride).
- NaOH solution (0.3 N sodium hydroxide).
- Specimens: Concrete cylinders (100 mm dia × 200 mm length) cut into 50 mm thick discs.

Stepwise Procedure:

1. Specimen Preparation

- Cast concrete cylinders (100 × 200 mm).
- Cure them for 28 days in water.
- Cut the cylinders into 50 mm thick discs using a diamond saw.
- Coat the curved surface with epoxy to prevent side penetration.

2. Conditioning of Specimens

- Dry the discs in an oven at 50 °C for 3 days.
- Place in a vacuum desiccator and apply vacuum for 3 hours.
- Saturate with 0.3 N NaOH solution under vacuum for 1 hour.

3. Test Setup

- Place specimen in the RCPT test cell.
- One side of disc exposed to 3% NaCl solution (chloride source).
- Other side exposed to 0.3 N NaOH solution.
- Apply 60 V DC across the specimen (NaCl side = cathode, NaOH side = anode).

4. Measurement

- Record the current (mA) immediately after applying voltage.
- Continue recording current at 30 min intervals for 6 hours.

5. Calculation

- The total charge passed (Q) is calculated by integrating the current vs. time curve. Units: Coulombs.
- $Q = 900 \times (I_0 + 2\ I_{30} + 2\ I_{60} + ... + I3_{60})$

  Where:
- I_0, I_{30},.....I_{360} = Current in Amperes at respective time intervals
- 900 = Conversion factor (30 min = 1800 sec; but since current recorded every 30 min, simplified factor used).

AGGREGATE TESTING IN LABORATORY

- Specific gravity and water absorption test
- Aggregate impact value test
- Aggregate abrasion value test

1. SPECIFIC GRAVITY AND WATER ABSORPTION TEST OF COARSE AGGREGATE

Apparatus Required

- Wire basket (2 mm mesh)
- Container for water
- Absorbent cloth
- Balance (0.5 g accuracy)
- Shallow tray
- Oven

Test Procedure

Sample Preparation

1. Take about 2 kg of dry coarse aggregate (retained on 10 mm IS sieve).
2. Wash to remove fines and dry in oven at **100–110°C** for 24 ± 4 hours.
3. Cool in an airtight container to room temperature.

Determination of Weights

1. **Weight of oven-dry sample in air (W1):**

   Weigh the dry aggregate sample (cooled to room temp).

2. **Soaking:**

   Place the sample in a wire basket and immerse in distilled water at 22–32°C.

   Keep completely immersed and remove entrapped air by gentle agitation.

   Maintain immersion for **24 ± 0.5 hours**.

3. **Weight of saturated sample in water (W2):**

   After 24 hours, weigh the basket with aggregates in water.

   Then weigh the empty basket in water.

   Subtract to get **W2** = weight of saturated aggregate in water.

4. **Weight of saturated surface dry sample in air (W3):**

   Take out the aggregates from water, wipe surface with a damp cloth (SSD condition).

   Weigh immediately in air → W3.

Calculations

1. **Specific Gravity (SSD basis):**

$$G_{SSD} = W1 / (W3 - W2)$$

2. **Apparent Specific Gravity:**

$$G_{app} = W1 / (W1 - W2)$$

3. **Water Absorption (%):**

Water Absorption = (W3−W1) / W1 ×100

2. AGGREGATE IMPACT VALUE TEST

Apparatus Required

- Impact testing machine (with cylindrical cup & hammer of 13.5–14.0 kg, drop 38 ± 2 mm).
- Cylindrical measure (75 mm dia × 50 mm depth).
- Tamping rod (10 mm dia × 230 mm long).
- Balance (0.1 g accuracy).
- IS Sieves: 12.5 mm, 10 mm, and 2.36 mm.

TEST PROCEDURE

Sample Preparation

1. Take about 3.75 kg of coarse aggregate.
2. Sieve through 12.5 mm IS sieve and retain on 10 mm IS sieve. (Only aggregates between 12.5 mm and 10 mm size are used).
3. Oven dry at 100–110 °C for 4 hours, cool to room temperature.

Test Steps

1. Fill the cylindrical measure in three layers, each tamped 25 times with tamping rod.
   Level the top and weigh → this gives test sample (about 350 g).
2. Place sample in impact test mould (cylindrical cup of machine). Tamp it gently 25 times again with tamping rod.
3. Fix the cup on the base of the impact testing machine.
4. Raise the hammer and allow it to drop freely from a height of 38 cm. Apply 15 blows at an interval of about 1 second.

5. Remove the sample, sieve through 2.36 mm IS sieve. Collect fraction passing 2.36 mm and weigh (W2). The fraction retained on sieve is weighed separately (W1).

Calculations

1. Total weight of sample (W) = W1 + W2
2. Aggregate Impact Value (AIV):

AIV = W2 / W × 100

where:

- W2 = weight of fines passing 2.36 mm sieve
- W = total weight of sample

3. AGGREGATE ABRASION VALUE TEST (Los Angeles Abrasion Test, IS: 2386 Part IV – 1963):

Apparatus Required

- Los Angeles Abrasion Testing Machine (hollow steel cylinder with removable cover, rotates at 30–33 rpm).
- Steel balls (abrasive charge, standard weight and number as per gradation).
- Balance (accuracy 0.1 g).
- IS sieves (sizes depend on grading chosen).
- Oven (100–110 °C).

Test Procedure

Sample Preparation

1. Take 5–10 kg of aggregate depending on the grading (see IS table).
2. Sieve and prepare aggregates in standard grading sizes (e.g., passing 12.5 mm and retained on 10 mm, etc.).

3. Wash and oven dry the sample at 100–110 °C, then cool to room temperature.

Test Steps

1. Weigh test sample (W1).
    - Take the required quantity of aggregate for the chosen grading.
2. Place aggregate and abrasive charge (steel balls) into the Los Angeles machine.
    - Number & weight of balls depend on grading (e.g., 12 balls for Grading A, etc.).
3. Rotate the machine for 500 revolutions at 30–33 rpm.
4. Remove material and sieve through 1.70 mm IS sieve.
    - Collect and weigh the material retained on 1.70 mm sieve (W2).

Calculation

Abrasion Value (%) = (W1−W2) / W1 × 100

where:

- W1 = original weight of oven-dry sample (g)
- W2 = weight of aggregate retained on 1.70 mm sieve after test (g)

CONCRETE MIX DESIGN (IS 10262:2019 method, compatible with IS 456:2000)

Steps

1. Data Collection

Before starting, collect all required data:

- **Grade of concrete** (e.g., M20, M25, M30, etc.)
- **Type of cement** (OPC, PPC, etc.) and its strength

- **Maximum size of aggregate** (10, 20, or 40 mm)
- **Workability requirement** (slump value)
- **Exposure condition** (mild, moderate, severe, etc.)
- **Water-cement ratio limits** (from IS 456)
- **Aggregate type & grading** (angular, rounded, etc.)
- **Admixtures** (if used)

2. Target Mean Strength

Calculate **target strength (fck'):**

$fck' = fck + 1.65 \times S$

where:

- fck = characteristic compressive strength at 28 days (e.g., 20 N/mm^2 for M20)
- S = standard deviation (from IS 10262 Table 8, usually 4–5 N/mm^2)

3. Selection of Water-Cement Ratio

- Use relationship between strength and w/c ratio (or refer to past test data).
- Ensure chosen w/c ratio $\leq$ maximum limit from IS 456 (durability requirements).

4. Selection of Water Content

- Choose water content from IS 10262 Table (based on max. aggregate size & slump).
- Correct water content for required workability & admixtures.

5. Calculation of Cement Content

Cement Content = Water Content / Water Cement Ratio

Check against IS 456 requirements (minimum cement for durability).

6. Proportion of Aggregates

- From IS 10262 tables, take volume proportions of coarse and fine aggregates depending on maximum size & zone of sand.

- Adjust based on workability and grading.

7. Calculation of Mix Proportions (by Weight)

Now calculate weights of all ingredients for 1 m^3 of concrete:

1. Calculate volume of concrete = 1 m^3
2. Calculate volume of cement, water, and admixture
3. Balance volume = volume of aggregates (fine + coarse)
4. Convert to weights using specific gravity.

8. Trial Mix Preparation

- Prepare a trial batch with calculated proportions.
- Check for workability (slump test) and strength.
- Adjust water, sand, or admixture as needed.

9. Final Mix Proportion

After trials, finalize mix proportions in the form:

Cement : Fine Aggregate : Coarse Aggregate : Water Cement

PLASTIC AGGREGATE PREPARATION

Collection of Waste Plastic

- Collect post-consumer plastic waste (polyethylene bags, PET bottles, polystyrene, polypropylene containers, etc.).
- Avoid PVC (Polyvinyl Chloride) since it releases toxic gases on heating.

Cleaning & Drying

- Wash the plastic waste to remove dirt, paper, and organic matter.
- Dry properly to prevent moisture content.

Shredding / Cutting

- Cut or shred the plastic waste into small pieces (5 mm – 20 mm size) using a shredding machine.
- Particle size can be adjusted depending on the required aggregate grading.

Melting / Heating

Two approaches are generally used:

a) Direct Heating Process

- Heat the shredded plastic in a controlled furnace (temperature: 160°C – 250°C, depending on plastic type).
- Plastic softens and becomes moldable.
- Avoid overheating to prevent toxic gas release.

b) Extrusion Process

- Feed the shredded plastic into an extruder machine.
- Extruder melts and converts it into long threads/rods.
- These threads are then cooled in water and cut into desired aggregate sizes.

Formation of Aggregates

- The molten or softened plastic is molded or pelletized into rounded shapes resembling natural aggregates.
- Sizes generally produced: 5 mm to 20 mm (similar to coarse aggregate).

Cooling & Hardening

- The hot plastic pellets/aggregates are cooled by air or water quenching.
- They harden into lightweight plastic aggregates.

Sieving & Grading

- The cooled aggregates are passed through IS sieves to obtain the required size fractions (10 mm, 12.5 mm, 20 mm, etc.).

Characteristics of Plastic Aggregates

- **Lightweight** (lower density than natural stone).
- **Non-absorbent** (nearly zero water absorption).
- **Durable** (resistant to chemical attack).
- **Smooth surface texture** (may affect bonding with cement paste).

STEPS FOR COATING OF *BACILLUS SUBTILIS* ON WASTE PLASTIC AGGREGATES

1. Preparation of Waste Plastic Aggregates

- Collect waste plastic (PET, HDPE, LDPE, etc.).
- Clean thoroughly with water and detergent to remove dirt and oil.
- Shred into coarse aggregate size (10–20 mm).
- Dry the plastic pieces at room temperature or in a hot air oven (40–50°C).

2. Bacterial Culture Preparation

- Take a strain of *Bacillus subtilis*.
- Grow in nutrient broth (peptone, beef extract, NaCl) under sterile conditions.
- Incubate at ~37°C for 24–48 hours to allow spore formation.
- Harvest bacterial suspension (optical density adjusted to ~10^7–10^8 cells/ml).

3. Preparation of Coating Solution

- Mix the bacterial suspension with a nutrient carrier to protect and attach bacteria, e.g.:
  - Calcium lactate (acts as a nutrient for $CaCO_3$ precipitation).
  - Sodium alginate (forms a gel-like coating for encapsulation).
- Stir until uniform solution is formed.

4. Coating Process

- Dip the shredded plastic aggregates into the bacterial suspension with carrier.
- Keep immersed for 15–30 minutes to allow bacterial adhesion.
- Remove the plastic aggregates and spread on sterile trays.
- Air-dry or oven-dry at low temperature (≤40°C) so that bacteria remain viable.

5. Curing & Storage

- Store coated aggregates in sealed containers at room temperature.
- Use coated aggregates within 24–48 hours for best bacterial activity.
- Alternatively, coat and dry them completely, then store for longer periods (spores remain dormant until activated in concrete).

6. Incorporation into Concrete

- Use bacterial-coated plastic aggregates in concrete mix design (partial replacement of natural aggregates).
- During hydration, bacteria remain dormant but activate when moisture penetrates cracks, precipitating $CaCO_3$ and improving bonding/self-healing

EXPERIMENTAL PROGRAM:

1. **Plastic Aggregate Preparation**
2. **Cultivation of Bacteria**
3. **Bacterial Coating**
4. **Mix Proportioning**
    - Control Mix (0% plastic).
    - 5%, 10%, 15% plastic aggregate replacement (with and without bacterial coating).
5. **Casting & Curing**
6. **Testing**

PREARATION OF PLASTIC AGGREGATE

- Plastics are collected from the disposal area and it is sorted out (PVC avoided)
- Then it is washed out to remove dirt, paper and organic matter
- Then it is surface dried.
- Then it is crushed into small pieces
- Then it is heated into furnace to make it soft and moldable
- Then molten or softened plastic is molded or pelletized into rounded shapes resembling natural aggregates, sizes generally produced: 5 mm to 20 mm
- The hot plastic pellets/aggregates are cooled by air or water quenching.
- The cooled aggregates are passed through IS sieves to obtain the required size fractions (10 mm, 12.5 mm, 20 mm, etc.).
- Following are the test results of plastic aggregates

Sr.No	Property	Average value
1	Density	0.81kg/m3
2	Impact value	4.54%
3	Abrasion value	9.54%
4	Specific gravity	1.06
5	Water absorption	5%
6	Melting point	75-100 C

CULTIVATION OF BACTERIA

- Prior to culturing of bacteria, the culture medium is prepared. A culture media consists of water, carbon source, nitrogen source, trace elements and other growth factors. For the media, the Nutrient broth is used with the compositions as mentioned in below table.
- The pure culture of Bacillus subtilis B. subtilis (MTCC.736), which facilitates the precipitation of calcium carbonate, was collected from the National chemical laboratory Pune, India and is put on a petri dish.
- To avoid contamination by other species, all glassware and instruments used were previously sterilized. The bacteria B. subtilis forms irregular dry white colonies on the petri dish.

- Using a platinum inoculating loop, few colonies, that is a loop full of required bacteria is transferred into the prepared culture medium in a conical flask as part of inoculation.
- This inoculated media is then kept for incubating at 37^{oC} temperature consisting of 150 rpm orbital shaker for overnight and a test of Optical Density was then done to determine the concentration.
- The bacterial concentration is kept at 108 cells/ml. The concentration of bacteria is determined by finding ''Optical density'', using a spectrophotometer. As visible light passes through a cell suspension the light is scattered. The amount of light scattered, measured in a spectrophotometer gives a value for Optical density. An OD of 1 shows a concentration of 0.8 _ 109 cells/ml.

Composition of nutrient broth. (for every 1000 ml of water).

Sr.No	Composition Content	Quantity
1	Peptone	5 g/l
2	Sodium chloride	5 g/l
3	Yeast extracts	2 g/l
4	Urea	20 g/l
5	Calcium chloride	15 g/l
6	NH4Cl	2 g/l

COATING OF BACILLUS SUBTILIS ON PLASTIC AGGREGATES

- Strain of Bacillus subtilis is taken
- It is grown in nutrient broth NaCl under sterile conditions
- It is incubated at ~37°C for 24–48 hours
- The bacterial suspension mixed with a nutrient carrier i.e. calcium lactate to protect and attach bacteria
- It is stirred until uniform solution is formed.
- Aggregates are dipped into the bacterial suspension with carrier.
- It is kept immersed for 15–30 minutes to allow bacterial adhesion.
- Then aggregates are removed and spread on sterile trays.
- Then these are air-dried so that bacteria remain viable.

CONCRETE MIX DESIGN

- Grade designation: M25.
- Type of cement: Ordinary Portland cement 53 grade
- Max. size of aggregate: 20mm
- Type of exposure: Moderate
- Minimum cement content M25: $300 kg/m^3$...as per IS 456:2000 for M25
- Max water cement ratio for M25: 0.50
- Max temperature for concrete at the time of placing: 30-32 degree Celsius
- Slump: 25-75mm
- Compaction factor: 0.85

Target strength: fck +1.65 x S = 25 + 1.65 x 4 = 31.6 N/mm^2

Selection of water cement ratio:

For moderate, exposure and maximum size of aggregate 20mm

Water cement ratio =0.6

Adopt Water cement ratio = 0.5 < 0.6 -------hence ok.

Maximum water content = 186 lit

Water content / cement content = 0.5

Cement content = water content / 0.5 = 186 / 0.5 = 372 kg/m^3

Volume of CA and FA (IS 10262-2009)

Volume of CA = 0.62

Volume of FA = 1 - 0.62 = 0.38

Mix calculation

Volume of concrete = 1m^3 ....(A)

Volume of cement =

Mass of cement / Mass density = 372 / (3.15x1000) = 0.118m^3 ....(B)

Volume of water =

Mass of water / Mass density = 186 / (1x1000) = 0.186m^3 ....(C)

Volume of all in aggregate = A-(B+C) =1-(0.118+0.186) = 0.696m^3 ....(D)

Mass of CA = Density x Sp. Gravity of CA x volume of CA x 1000

= 0.696 x 2.81 x 0.62 x 1000 = 1212.57kg

Mass of FA= Density x Sp. Gravity of FA x volume of CA x 1000

= 0.696 x 2.33 x 0.38 x 1000 = 629.46kg

Cement	FA	CA	Water
372kg	629.46kg	1212.57kg	186lit
1	1.69	3.25	0.5

Ratio = 1:1.69:3.25

Volume of one cube = L X B X H

Materials calculations:

Ratio = 1:1.69:3.25 = 5.94

Volume of one cube = 150 x 150 x 150 = 0.00337 mm^3

Dry volume extra 52% of wet volume

Dry volume = 52/100 x 0.00337 = 0.00513

Quantity of Cement = 1/5.94 x 0.00513 = 0.000863m^3 = 0.025 bags = 1.25 kg.

Quantity of Sand = 0.000863 x 1.69 = 0.00145m3= 1600x0.00145 = 2.32kg.

Quantity of Aggregate = 0.000863 x 3.25 = 0.00280m^3 = 1600x0.0028 = 4.82kg.

Quantity of Water: = w/c - 0.5 = W = 0.5 x C = 0.5x1.25 = 0.625 kg.

Cement	1.25kg
Sand	2.32kg
Aggregate	4.82kg
Water	625ml

Quantity of Shredded Plastic for Different %

5% Aggregate Replacement -

Shredded plastic aggregate = 4.82kg x 5% = 0.241kg = 241.9gm.

Coarse aggregate = 4.579kg.

10% Aggregate Replacement -

Shredded plastic aggregate = 4.82 x 10% = 0.482kg = 489.2gm.

Coarse aggregate = 4.338kg.

15% Aggregate Replacement -

Shredded plastic aggregate = 4.82 x 15% = 0.723kg = 723.9gm.

Coarse aggregate = 4.097kg.

20% Aggregate Replacement -

Shredded plastic aggregate = 4.82 x 20% = 0.964kg = 964.9gm.

Coarse aggregate = 3.85kg.

CASTING ,CURING & TESTING

Materials Required

- Cement (OPC 53 grade)
- Fine aggregate (river sand, clean and dry)
- Coarse aggregate (natural + waste plastic aggregates + waste plastic aggregates with bacterial coating)
- Water (potable quality)
- Bacterial culture
- Measuring equipment (weighing machine, measuring cylinders, trowel, tamping rod)
- Cube moulds (150 mm × 150 mm × 150 mm)
- Vibrating table or tamping rod

We have adopted the following steps for Casting, Curing and Testing

Step 1: Mix Proportioning

- We have adopted the mix design proportions finalized (1:1.69:3.25 with a w/c ratio of 0.5).
- Weighed all materials accurately according to the designed batch.
- For bacterial concrete, we have used the bacterial-coated plastic aggregates

Step 2: Mixing of Concrete

- We have cleaned the mixing pan.
- Mixed the cement and fine aggregate thoroughly in dry condition until uniform color is obtained.
- Added coarse aggregates (both natural and plastic aggregates) and mixed again.
- Gradually added water while mixing continuously to obtain a uniform consistency.
- Continued mixing for about 3–5 minutes until a homogeneous mixture is achieved

Step 3: Workability Test

- We have performed the slump test immediately after mixing to check workability.
- Recorded the slump value for each batch (important for comparison between mixes)

Step 4: Preparation of Moulds

- We have cleaned and oiled the inner surfaces of cube moulds properly.
- Assembled the moulds tightly with all bolts and nuts fixed securely.

Step 5: Filling and Compaction

- We have filled each cube mould in three equal layers (approximately 50 mm thick per layer).
- Compacted each layer by tamping rod 25 blows per layer and also with vibrating table (We have avoided over-vibration).
- Leveled the top surface with a trowel for a smooth finish.
- Labeled each cube clearly (e.g., Mix A – Control, Mix B – Plastic Aggregate, Mix C – Bacterial Plastic).

Step 6: Curing

- We have kept the moulds undisturbed for 24 hours at room temperature (covered with damp cloth or plastic sheet).
- After 24 hours, demoulded the cubes carefully.
- Placed the cubes in a curing tank filled with clean water maintained at 27 ± 2°C.
- Cured for the required duration — typically 7, 14, and 28 days.

Step 7: Testing

- After curing we have removed cubes and wiped off excess water.
- Tested the cubes for compressive strength using a compression testing machine (CTM).
- Recorded the failure load and calculated compressive strength using:

  $\mathbf{f_c = P / A}$

  Where,

  f_c = Compressive strength (N/mm^2)

  P = Maximum load at failure (N)

  A = Cross-sectional area of cube (mm^2)

Step 8: Observations & Reporting

- We have recorded observations for each mix (slump value, appearance, load at failure, compressive strength).
- Compared the results of control, plastic aggregate, and bacterial-coated plastic aggregate concrete.
- Noted differences in strength and failure pattern.

RESULTS

NATURAL AGGREGATE - Mix A					
Sr. No.	**% replacement**	**Days**	**Load KN**	**Strength (Mpa)**	**Average**
1	0	7	427	18.98	19.25
			433.1	19.25	
			439.2	19.52	
		14	518.5	233.04	22.76
			506.3	22.50	
			512.4	22.77	
		28	580	25.78	26.44
			610	27.11	
			595	26.44	

REPLACEMENT WITH PLASTIC AGGREGATE - Mix B					
Sr. No.	**% replacement**	**Days**	**Load KN**	**Strength (Mpa)**	**Average**
1	5	7	441	19.6	19.89

			447.3	19.88	
			453.6	20.16	
		14	535.5	23.8	23.51
			522.9	23.24	
			529.2	23.52	
		28	607.5	27	27
			630	28	
			618.75	27.5	
2	10	7	401.63	17.85	18.11
			407.36	18.02	
			413.10	18.36	
		14	487.69	21.68	21.43
			476.21	21.17	
			481.95	21.42	
		28	551.25	24.5	25
			573.75	25.5	
			562.5	25	
3	15	7	371.7	16.52	16.76
			377.01	16.76	
			382.32	16.99	
		14	451.35	20.06	19.81
			440.73	19.59	
			446.04	19.82	
		28	499.5	22.2	23
			531	23.60	
			522	23.2	

4	20	7	326.03	14.49	14.7
			330.68	14.69	
			335.34	14.91	
		14	395.89	17.59	17.39
			386.57	17.18	
			391.23	17.39	
		28	445.5	19.8	20.5
			465.75	20.7	
			472.5	21	

PLASTIC AGGREGATE WITH BACTERIA - Mix C

Sr. No.	% replacement	Days	Load KN	Strength (Mpa)	Average
1	5	7	461.3	20.50	20.80
			467.89	20.80	
			474.48	21.09	
		14	560.15	24.90	24.61
			546.97	24.31	
			553.56	24.60	
		28	645.75	28.7	29.09
			659	29.29	
			658.75	29.28	
2	10	7	417.38	18.55	18.82
			423.34	18.82	
			429.30	19.08	

		14	506.81	22.53	22.24
			494.89	21.99	
			500.85	22.26	
		28	585	26	26.5
			596.25	26.5	
			607.5	27	
3	15	7	378	16.8	17.05
			383.4	17.04	
			388.8	17.28	
		14	459	20.4	20.15
			448.2	19.92	
			453.6	20.16	
		28	528.75	23.5	24
			540	24	
			551.25	24.5	
4	20	7	338.63	15.05	15.29
			343.46	15.27	
			348.3	15.48	
		14	411.19	14.28	18.07
			401.51	17.85	
			406.35	18.06	
		28	472.5	21	21.5
			438.75	21.5	
			495	22	

CONCLUSIONS

1. Conventional Concrete (Natural Coarse Aggregate)

1. The control mix (0% replacement) achieved an average compressive strength of 26.44 MPa, satisfying the target strength for M25 grade concrete.
2. The results exhibited uniformity with low variability, indicating consistent material quality and proper curing.
3. This serves as the baseline for evaluating the effect of plastic and bacterial-coated plastic aggregates.

2. Concrete with Waste Plastic Aggregate (Uncoated)

- Partial replacement of natural coarse aggregates with waste plastic aggregates was done at 5%, 10%, 15%, and 20% levels.
- The maximum compressive strength was observed at 5% replacement (27 MPa), which is slightly higher than the control mix.
- Beyond 5%, the strength gradually decreased:
  - 10% replacement: 25.00 MPa
  - 15% replacement: 23.00 MPa
  - 20% replacement: 20.50 MPa
- The reduction in strength at higher replacements is attributed to the smooth surface texture, non-polar nature, and poor interfacial bonding of plastic with the cement matrix.
- Hence, up to 5–10% replacement of plastic aggregate can be considered feasible for non-structural or light concrete applications.

3. **Concrete with Bacteria-Coated Plastic Aggregate**

- When the waste plastic aggregates were coated with Bacillus subtilis bacteria before mixing, the compressive strength improved notably at all replacement levels.
- The average 28-day strengths were:
- 5% replacement: 29.09 MPa
- 10% replacement: 26.5 MPa
- 15% replacement: 24 MPa
- 20% replacement: 21.5 MPa
- The bacterial coating enhanced bonding between plastic and cement paste due to microbial-induced calcium carbonate ($CaCO_3$) precipitation, which filled micro-voids and roughened the plastic surface.
- The improvement was most significant at 5% replacement, where strength increased by about 11.7% over the control mix and 7.4% over uncoated plastic concrete.

4. Overall Findings

- Optimum compressive strength is achieved with 5% replacement of coarse aggregate by bacteria-coated plastic aggregate.
- The bacterial coating mitigates the strength loss caused by plastic addition and contributes to improved interfacial transition zone (ITZ) quality.
- The use of bacteria-coated plastic aggregates thus represents a sustainable, eco-friendly, and strength-retaining alternative to conventional concrete by:
  - Reducing plastic waste,
  - Lowering environmental impact
  - Enhancing concrete performance.

Final Conclusion

Among all mixes tested, concrete containing **5% bacteria-coated plastic aggregate** exhibited the best balance of strength, sustainability, and performance. The bacterial treatment significantly improved bonding and compressive strength compared to uncoated plastic aggregate mixes, demonstrating its potential for sustainable construction applications.

NOTATION

The following symbols are used in this technical note:

G = Specific gravity

ρ = Density

L = Length of concrete mould

B = Width of concrete mould

H = Height of concrete mould

fck = Characteristic compressive strength

fck' = Target compressive strength

S = Standard Deviation

$S28$ = 28 days compressive strength of concrete in N/mm2

REFERENCES

1. Gehad A. M. Metwally, Mohamed Mahdy, Ahmed H. A. El-Raheen, 'Performance of Bio Concrete by using Bacillus Pasteurli Bacteria' Civil Engineering Journal, Vol. 6, No. 8, August 2020.
2. Azad Khajuria and Puneet Sharma 'Use of Plastic Aggregates in Concrete' International Journal of Innovative Technology and Exploring Engineering (IJITEE) ISSN: 2278-3075 Volume -9 issue -1, November 2019
3. M Mahesh, B Venkat Narsimha Rao, CH. Satya Sri 'Re-use of Polyethylene Plastic Waste in Concrete' International Journal of Engineering Development and Research; Volume 4, Issue 4 ISSN:2321-9939.
4. Nain, Nidhi, R. Surabhi, N.V. Yathish, V. Krishnamurthy, T. Deepa, and Seema Tharannum. 'Enhancement in Strength Parameters of Concrete by Application of Bacillus Bacteria'Construction and Building Materials 202 (March 2019
5. IS 456. 2000. *Indian standard code of practice for plain and reinforced concrete*, Bureau of Indian Standards, New Delhi, India
6. Text book by M.S.Shetty "Concrete Technilogy"

FSC
www.fsc.org
MIX
Papier | Fördert gute Waldnutzung
FSC® C083411

Kolibri 360 GmbH
Ferdinand-Jühlke-Straße 7
99095 Erfurt, Deutschland
produktsicherheit@kolibri360.de

Druck:
CPI Druckdienstleistungen GmbH
im Auftrag der
Kolibri 360 GmbH
Ferdinand-Jühlke-Str. 7
99095 Erfurt